C000292048

TANGLED IN VOW & BESEECH

poems

Jill McCabe Johnson

MoonPath Press

Copyright © 2024 Jill McCabe Johnson.
All rights reserved.

No part of this publication may be reproduced, distributed, or transmitted in any form or by any means whatsoever without written permission from the publisher, except in the case of brief excerpts for critical reviews and articles. All inquiries should be addressed to MoonPath Press.

Poetry
ISBN 978-1-936657-83-4

Cover art: Barred Owl by Mshake via iStock

Book design: Tonya Namura, using
Times New Roman (text) and Mr Eaves (display).

.

MoonPath Press, an imprint of Concrete Wolf Poetry Series, is dedicated to publishing the finest poets living in the U.S. Pacific Northwest.

MoonPath Press
c/o Concrete Wolf
PO Box 2220
Newport, OR 97365-0163

MoonPathPress@gmail.com

http://MoonPathPress.com

for Jeff

ACKNOWLEDGMENTS

Barzakh Magazine
“For Immediate Release”
“Poem in Which I Call Out Fellow Poets for Painting
Abuse as Sexy”
“Trumpeters’ Song”

Bracken
“Starlight Tour”

Cascadia Field Guide: Art, Ecology, Poetry
“Tadpoles”

Crab Creek Review
“Boxed In”

Cultural Daily: Poets on Craft
“Ars Poetica”

Diode Poetry Journal
“Thoughts and Prayers”
“Gaslight”

I Sing the Salmon Home Anthology
“Bedded in Wet Rock Like Any Other Roe”

Inflectionist Review
“Love’s Blind Contour”

Jake: The Anti-Literary Magazine
“Embodying Language” (Best of the Net nominee)
“Under the Thumb of Patriarchy” (Best of the Net nominee)

Madrona Project
“Origin Story—Great Blue Heron”

More In Time: A Tribute to Ted Kooser
“1,001 Things to Amend Before You Die—Excerpt 244-258”

Ōde
“In this land, everything” (Pushcart nominee)

One Art
"Stargazing"

Page & Spine
"Breakaway"
"Raptor"

Passengers Journal
"On the Bus"

Terrain.org
"Dear World,"

The Tishman Review
"Packing for Peace" (Finalist in the Edna St. Vincent Millay Poetry Prize)

Waxwing
"Slipsilver"
"When You First Came Home"

The following poems, some revised, appear in the chapbook *Pendulum* (Seven Kitchens, 2018), finalist in the Rane Arroyo Prize in Poetry: "On the anniversary of the first time we kissed," "Travel Journal," "Apnea," "Apogee of Apathy," "A Quantum Theory of Love," "To My Brother on the Loss of his Wife," "For Campbell, Jumping Out the Car Window," and "A Boreal Cure."

GRATITUDE

Thank you to poet and publisher, Lana Hechtman Ayers, and designer extraordinaire, Tonya Namura, for the many poetry books you've brought into the world, including this one. Your tireless, generous support of poetry is a gift to poets and readers alike.

I am incredibly fortunate and deeply grateful for the feedback, encouragements, and kindnesses my friends and fellow writers gave as I worked on these poems the last several years, especially Susan Kim Campbell, Gail Folkins, Jami Macarty, Julie Riddle, Tina Schumann, Ana Maria Spagna, Joannie Stangeland, the Poets on Pinneo, and the San Juan Island Winter Poets. Thank you for your friendship.

I'm grateful, too, to the following organizations who have provided support in the writing of these poems:

Artist Trust
Atelier de la Rose, Montcabrier, France
Brush Creek Foundation for the Arts
Flying Squirrel Writing Residency
Hedgebrook
Helen Riaboff Whiteley Center
The National Endowment for the Humanities
Playa Artist Residency
Vashon Artist Residency

Your support for the arts, humanities, and sciences makes the world a better place.

To my son, Jeff, to whom this book is dedicated, thank you for your brilliance, insights, integrity, playful humor, and generosity of spirit. You inspire me and make me laugh in equal measure.

And finally, to my husband, my life-mate, my partner, and my joy: an unending river of love and thanks. You and your entire family have been an unexpected gift of love and graciousness in my life. How did I ever get so lucky?

CONTENTS

TANGLED IN VOW & BESEECH

EMBEDDED
IN BONE & BROKEN PROMISES

PACKING FOR PEACE

after Matt Hohner's "How to Unpack a Bomb Vest"

when you say you're packing   fill your bags
armed   with dandelion seeds
carrying   wishes you can launch
I want to believe   in the wind—cram feathers
you're collecting   and the taste of
enchantments   wild strawberries
a kind of magical mettle   into pockets
stuffing fear   of torn memories
with friendship   palm the scent of lilac
maybe a nest of   pinesap, spikemoss,
hummingbird eggs   and hollowed out cedar
filled with marvels   offer these to strangers
peace requires more than   unspent
weapons and   promises the
explosive   blossoming
shells   of dreams

THOUGHTS & PRAYERS

On behalf of the people for the people we'd like to express
our deepest condolences. For the victims and the families
of the victims we send our thoughts and prayers.
Thoughts & prayers to the victims and the families
after such tragic loss we'd like to express on behalf of our
deepest we send condolences in this moment of grief.
In this moment our thoughts & prayers are with you
all, sending to the students for their heartbreaking
losses the victims & their families our deepest
expression on behalf of the beautiful souls in this difficult
may they rest time in peace those felled by another mass
moment of silence and in our grief we wanted you to know
we're pulling for you stay strong may your memories
comfort you we'll light a candle you'll always be
in our hearts we'll never forget the victims and their
families we'd like to express our deepest for your pain
and prayers may they rest in heartbreaking moments
as we mourn these losses to unnecessary so unnecessary
violence only the good our nation die young & we grieve
with you you'll always be you'll never be forgotten
passed away on earth as it is in heaven give us our daily
thoughts & prayers on behalf of victims & their loved ones
for the bereaved the survivors expecting changes legislation
it's too soon too disrespectful guns don't kill people
people kill people so take care take cover arm yourselves
we're rooting for you we're sorry so very sorry
in the name of the father & the son & the holy
gun peace be with you & also with you forever & ever
Amen

STARLIGHT TOUR

The practice of police officers driving... an Aboriginal man/youth to the outskirts of town during the night, and often during winter, and leaving them there... Neil Stonechild, a Native youth froze to death in 1990, after being taken on a 'starlight tour' by two Saskatoon police officers.

—Urban Dictionary

How bright and dazzling
on a Saskatchewan night
like polished badges
you could almost warm
beneath each cold waning
even your breath
an iced mist
of the north
removed
subtracted
the open tundra
stretched out
under polar winds
with two red stars
receding down the
highway two
devil hearts
and you seventeen
Ojibwe Saulteau
glimmering like a stone

the stars
shimmering
the winter air so clear
your cuffed hands
fire
spreading
a Kristallnacht
a pogrom of one
separated
from home
the plain
like a bed
wrapping like a blanket
of rime
yellowhead
eyes of the wolf
two taillights forsaking
an original
abandoned
child

BOXED IN

If I arranged the letters carefully | a little tighter here | a little wider there | and justified left to right | without gap or division or polarization | if I listened to FOX and Sinclair | as readily as NPR and BBC | if I typed with an eye | toward balance | maybe each poem could carve a window | or box | shaped unilaterally | not knuckled like a fist | or pointed like a finger | accusation | AK-47 | the rhetoric of *Boom* | if my hands could wing open | birds of blossom | articulated dreams | arrayed like a fan | matchbook incendiary | rustle of ignition | or if I sat on my hands | abated the click | the kneejerk | hot take | sweltering angry urge | and folded logic and reason and eyeroll and smdh in on themselves | a neat envelope shaped | like a ballot | notched with checks and chads | and signature fashioned | to match the original | if I counted as a person | true & self-evident | like all men | created equal | if my voice | my vote | tallied into something | other than rage | if I dug in | posts and erected | fencing to the north, east, west, south | NEWS | measured and fact-checked | not a map of destiny | so much as a chart | an atlas of steering | to guide us away from | our cartography of corruption | if I drew four lines | penned them in | on rough scratch paper | or an empty slate | a field for tilling | for sowing, reaping | where unfettered winds | cry in passing | paper like a sheet | stretched over a bed | smooth, freshly laundered | not like money | not like real estate deeds | traded for illicit cash | no, a bed for resting | for sleep | if we dare | wide and inviting | and isn't that the real seduction | to close our eyes | turn our backs and succumb | to slumber, to dreams | if I woke and stayed woke | if I snorted Adderall | or whatever it takes to stay alert | to courts to judges to FTC, EPA, DOJ, USPS | if I stopped SMDH | on Twitter and Facebook | and made calls, wrote letters | to crooks in lesser offices | *we're watching you* | if I put down my protest signs | and picked up a camera | in many languages

| camera means chamber | if I joined the panopticon | & posted viral videos | of politicians compromised | of nominees in compromising | positions | abuses of power | if I linked arms | or laid down arms | reached out | stood strong with neighbors | crossed my heart & swore to never ever | ever | give up

DEAR WORLD,

Good news! I found your missing socks
on the roadway today, tortured
with mud, trampled underfoot,
like a lost-and-found where loved ones
can rescue old friends
                              who'd marched with them
and protected them from the brutality
of winter's chill. Two friends fell in the path
of xenophobia—
                              that tired, old retread—
splayed flat and torn open like so much roadkill.
One got rescued by the neighbor's retriever,
who clutched him tenderly in his mouth,
an intimacy
                    most of us would never risk.
Another got pecked to death by shrikes.
The neighbors gazed on,
                              but no one came to her defense.
What was it Elie Wiesel said about the perils
of indifference being always the friend
of the enemy?
                    Goebbels, that Monster of Propaganda,
advised us to accuse our foes of the crimes
we ourselves are guilty of. Indifference
being among the worst of these. I found your socks,
Dear World.
                    I've placed each one delicately
in my mouth, where accusations typically reside.
They are here, waiting. Black and blue.
Yellow and red. If neglect is a kind of prison,
retrieval is a kind of grace.
                              Whenever you're ready.

TRUMPETERS' SONNET

outlined against the horizon,
a golden eagle
hunts atop wintry thermals
a black ode carving itself
along the tyranny of sky
a sidesweep of wind
its cloak of false hope and

America, you unmoor me
your democratic mirage
your tweets and caws
when the clouds blitz
heat hemorrhages
and morning spreads
fleeting diamonds

THUNDERCRACK

The shot resounded like ballpeen on sheet metal
and riffled us awake, bed covers dragged
over heads, our parachute of sleepy denial.

I found myself counting seconds between flash and echo,
like cultures who measure the lengths between knots,
a palpable mathematics of event and aftermath,

jab and bruise, gunshot and bullet, embedded
in bone and broken promises. Like this world
I had pictured so trustingly as a teen, walking home

on empty streets in the drowsy hour past midnight,
a gun popped in the distance, too far to hear
the steely ring of casings but I sensed

the way the air bent close to my ear with each
whistling missive. And who knows what hunters lurk.
Summer night, girl walking.

Did he aim to miss? Or did he nail a bullseye
that reverberates decades later, even from this marrow
of slumber, a memory I cannot purge. It hides

in the very amygdala that once was a moving target
on a night as safe as tonight.

APOCRYPHA

Funny how I used to believe the story of Adam and Eve
As a proverb to remind us to care for this world and each
Living being. A parable teaching us to honor our fathers,
Listen and obey. But the Bible's story of Eden and the Fall
Of Man was never about the fall and always about
Foisting blame: on God, the apple, woman, the snake.

Myths like the Abrahamic creation tale use the snake
As an obvious metaphor that all but the most naive
Notice. If you're still unsure what that's about,
Find a banana, popsicle, rocket, or gun. Even preachers
Allude at times to that same image, the phallus,
Linking the rod with the fallacy of an all-protective father.

Like in the case of Brock Turner and how his father
Offered a statement that his son had barely eaten his snacks
Following his arrest, and Dad claimed witnesses falsely
Maligned Brock—though he was seen raping and leaving
A woman half-naked behind a dumpster. Each
Nebulous excuse implied consensual sex gone awry and

Foolish acts. *That is steep price to pay for 20 minutes of*
Action, he told the judge. I imagine most fathers
Lie sometimes to protect their sons, but each
Lie points a finger at the victim or the errant snake.
One of the oldest rape defenses asks us to grieve
For the man whose libido is somehow his downfall.

Mirroring the argument that God's trap forced the fall of
Adam and Eve is another, and let's not forget about
Nominee Kavanaugh's defense that Dr. Kristine Blasey
Ford had falsely accused him. Did the other women falsely
Accuse, too? C'mon. Who is the snake? Who is the snake?
Legislators surely weren't fooled by his angry speech

Lacking any compassion as Kavanaugh screeched
Outbursts of forced histrionics and false
Fury. But the Senate's response affects us all. The stakes
Magnify, and Supreme Court or not, rape cases reward the
Adams of the world bellowing with a toddler's
Noisy tantrum, crying *It was entrapment—No, the snake—*

No, Eve—not about the pain inflicted
Or how each conspiring phallus
Would never betray father or brother. Only Eve, only Eve.

NIGHT VIGIL

O god of lost prayers
god of no promises
god of taciturn skies
where will you be
when the body resolves
and the exhale goes
searching will you
claim in your omni-
science you voiced
no assurances but
told us no lies?

GASLIGHT

We all know the silencing of women.
Like Brigid, the goddess the poets adored.

She invented keening for her son who died
on the battlefield. In honor of the healer and smith,

women tended a perpetual flame that cut through
darkness like a relay of whistles in the night.

Her oxen, Fe and Men, worked the fields in service
to her feminine powers. But medieval ministers

demoted Brigid from goddess to saint, stole
the fire of Imbolc when villagers lit winter wicks,

and renamed the festival Candlemass, a celebration
of Christian patriarchy and the false belief
in virginity as proof of purity.

Every woman whose original ideas
have been appropriated by men
and has been told it was a man's design

in the first place, carries the brilliance of Brigid
and the burden of the earliest man who refused
to shoulder responsibility for his own weakness.

The festival of Proserpine, the Eleusinian mysteries
of Demeter and Persephone, the goddess Februa—
each event venerated the dawn of ovulation
following the passage of menses.

It’s no wonder patriarchy rewrote
women’s cycles into shame,
silencing even our deepest
tides of feminine might.

ON THE BUS

Was it violence on the bus
when the drunken dude
unzipped to his tan lines
and bent a double exposure
close to my sleeping
12-year-old face?
                                        Was it violence
when his friends laughed
and goaded him to inch closer
fart in her face   shove in hard
so she can't breathe?
                                            Was it violence
when the guy stepped back
to reposition and my one opened
eye caught both eyes
of the driver who stared back
at the highway?
                                      Was it violence
for my girlfriends across the aisle
to stare straight ahead, too,
pretending they saw nothing,
didn't know me, were just
minding their own?
                                          Was it violence
when I pretended that faith
in humanity was my only
loss,
        my only injury?

POEM IN WHICH I CALL OUT FELLOW POETS FOR PAINTING ABUSE AS SEXY

I was almost seduced by wet metaphors
with thick drops of thumb-smudged honey
across a poetic line
    of swollen, half-parted lips

and how every woman who ever tripped
down your stairstep of stanzas had a voice
dry as a martini
    sipped by firelight, but

now spilled across her hunger torn open
and those sultry lids, that gaze plunged deep
in the flesh of rough-
    cut, over-ripe melon.

Yes. Almost. Though excuse me while I clear
my throat, while I push aside the ever-present
imprint and sting,
    the stain of the blow you bear

from her kiss, her glance, her unrelenting
depiction by you as still marred, shaken and sexy
as the slick of ink
    pouring from your pen

after her lover's last thrashing
while you—our somehow bruised poet—
sit soused
    in echo and ache.

C-WORD

The word hurled
like a four-pointed shuriken
aimed at soft tissue and meant
as distraction in war.

An open gash that bleeds.
Sometimes they call it a spurge,
mistaking its milky euphoria
for the binding sap of euphorbia.

Sometimes they shear the tender nub
where the nerve ends
coalesce into confluence.
To burn and to shrive.

Many fear it, the mouth of Smilodon,
saber-toothed and gaping
yet swath-like, silken.
Roof ridged, a ravenous maw.

Yet impenetrable, like a cave
whose depths know
the secret memories
before there ever was light.

1,001
THINGS TO AMEND

ADHESIVE CAPSULITIS

Strands radiate from the shoulder a temporary
        web of paralysis—sinews of pain—live wires
where memory rides like the fiery kindle on a fuse

        or the fast singe of hair hovering close
over the candle as I lean to lift the wine glass
        to toast our love and who knew it would blaze

so fast so hot so steady a zeal of potassium nitrate
        sparklers shooting glints of metal as blistering
as the jolts shooting down my arm now

        across my torso up my neck and you ask
are you all right are you in pain is there anything
        I can do but I raise my glass with the other arm

and tell you don't worry the hurt will subside
        this shoulder this life this magnificent forever
let it burn let it burn my darling my everything let it burn

NOTRE DAME

... experts say the beams, many of them dating to the cathedral's construction in the 12th and 13th centuries, became tinder-dry as they aged. ... Settled dust and debris could have made the area even more flammable.

—*The New York Times*

Nearly 1,700 priests and other clergy members that the Roman Catholic Church considers credibly accused of child sexual abuse are living under the radar with little to no oversight from religious authorities or law enforcement.

—Claudia Lauer and Meghan Hoyer,
Associated Press

We don't have a culture of accountability.

—Jesuit Fr. Hans Zollner

It starts as a flick, a spark
not of hate but indifference
burrowed into centuries
of dust and seasoned wood,
vestments ripped with mutinous

shame. It starts in the altar, the belfry,
the pew. It kneels before no one.
Towers of smoke, an apocalypse
rising, ash on the forehead, ash
under the tongue, a communion,

a catechism, a retribution, this cleansing
torrent of combustion, sacristy
of soot, hymn from spire.
Icons turned effigy, vestiges aloft,
buttressed and flying.

Rib vault changed to charcoal,
water to wine. Milagros ablaze
like a firebrand in Christ's brimstone
heart—beating, thunderous,
eternally aflame.

1,001 THINGS TO AMEND BEFORE YOU DIE — EXCERPT 244-258

Age Nine

244. Losing the quartz rock your oldest brother gave you to start your own collection
245. Not admitting you didn't know what it meant, and congratulating your brother when he told you he'd been laid-off from Boeing
246. Your rush of relief when you realized your parents hadn't called you downstairs to yell at you for jumping on the sofa
247. Realizing that while you'd been shouting and jumping on the sofa your parents were figuring out how to tell you your brother, their eldest son, had died from a seizure in his sleep
248. Not hugging your father as he trembled, swallowed back the impending landslide, then told you your brother was in God's hands now
249. Letting your new best friend borrow your brother's portfolio of sketches
250. Not tracking down your former new best friend when she moved away a few weeks later
251. Scarfing handfuls of chocolate chips, Cap'n Crunch, and potato chips for dinner while both your parents lay bedridden with pneumonia the following winter
252. Keeping, then not keeping, your mother's secret when she told you, decades later, that your brother's death may have been a suicide—relief from the constant storm of seizures
253. Believing somehow that your misplaced congratulations had contributed to his sense of isolation and that you were partially responsible for his death

254. Deluding yourself into thinking his soul goes with you everywhere
255. Knowing his absence has become a bigger presence in your life than your handful of thread-worn memories
256. Being unable to distinguish between missing him and the pang conjured each time you think of missing him
257. Pretending you see his face in every rock
258. Your inability to see his face in every rock

MAY DAY

We leapt from the slow-moving pickup
        and snatched stems by the handful,
tulip, hyacinth, narcissus,

from garden beds and landscaped borders.
        Young hoodlums, we fancied ourselves
more Robin Hood than wretched vandal.

Drunk on stolen beer, we drank
        the pilfered perfume of blossoms heaped
into their new bed. Like our unsuspecting

friends, deep in slumber as we tip-toed
        up their paths and onto their porches,
delivering great armloads of flowers

heavy with dirt on uprooted bulbs.
        Dumb teens, we couldn't imagine
loss till we saw how neighbors woke

to gardens trampled and torn,
        the one bit of beauty
they'd waited for all winter.

ELEGY FOR TWO BROTHERS

She softened chopped onions in a slow sizzle of butter—
by then the allium-induced tears and all our childhood
woes long forgotten as the caramel sang
to marjoram, oregano, bay leaf, and thyme.

Meat browned as tomato simmered into an afternoon nap,
the day's dandelions gathered, trees climbed, toes stubbed.
Once, a dream of a youngest brother—trapeze artist,
tightrope walker—balanced over the abyss.

Once, a conjuring of webbed leaves, maple and chestnut,
we layered and fluttered like Egyptian fans. Cleopatra
and her entourage back from the neighbor's orchard,
where we plucked plums, pale yellow, understory purple.

Once, a race with a dinosaur who reared up,
arms a menace, and chased us like an oldest brother
through prehistoric jungles, across the steppes
and savannah of our own backyard.

I could tell you of the time my mother lifted the boiling
pot of spaghetti, leaving behind a coiled orange snake
that lured both my hands to warm themselves,
handplant of flesh to fresh heat.

Or the knife she used to slice into mushrooms, garlic,
and her own thumb so many times she called the blade
her scalpel. We joked dinner was Communion.
Eat of my body. Drink of my blood.

But no, I won't tell you of the Tyrannosaurus Rex who fell,
when he stopped his seizure meds, or the Velociraptor
who, too late, chose salad over heaped platters of spaghetti
and bread slathered in butter. Two brothers

who drifted, dreamed and drank of the elixir of ever-after,
who nestle in that long, luscious, luxuriant nap, even still.

APNEA

for Mark

Was it you who showed me how an inverted cup can snuff a candle, that flame requires oxygen to breathe?

When morning banked silence in your corner pocket, I remembered how you would lean your weight against your elbows, each wrist pressed into metal, fingertips flicking the corresponding pinball paddles, as bells clanged and points racked. You nudged the steel balls into extra rounds, twitched the angled table right to the edge of *Tilt* and *Game Over* while the rest of us waited, sometimes hours, to take our turn.

As an adult, you smoked cigars, drank Jägermeister, always on the hunt for your next indulgence, next pleasure. Moderation an anathema, and we swore this burning at both ends business would do you in. How surprised we all were that last Christmas to see you fifty-some pounds lighter, cheekbones emerging, mottled but spry.

It won't surprise you the relief we felt when you snubbed the ember of your last cigar.

Remember reading comic books by flashlight, musky tent, how we knew even then we couldn't last forever? "I'm going down in a fiery crash," one of us said, but you wanted to die in your sleep.

Apnea comes from the Greek, an interrupted breathing.

At night, I try not to picture how flesh, suddenly relieved of fat will loll and sag on itself, how a body sleeping can slump like a wet bag draped over unwitting bones.

When we were kids, I could sit for hours watching the flames on the patio bonfire while you slow-roasted marshmallows on a sharp whittled stick.

I always loved that old Swedish *klockspel* carousel with the brass fan that spun from the candles' heat. Angels rang the bells with each turn, and Gabriel lifted his trumpet on high.

Every Christmas you brought cigars to share, and each year I lit one. Bonded in breath and flame.

I can still see your hand lowering the metal cone over the flame of the candle, can still hear your voice explaining how fire needs air, still watch the last wisp of smoke wicking away.

TO MY BROTHER ON THE DEATH OF HIS WIFE

You are dangling
an unpredictable series
of geometries
the trajectory of grief's complications
mind-boggling and exquisite
once two seeds
mirrored side-to-side
woodwings protruding
but now severed
let yourself float
even in plummet you remain
a tadpole suspended in spiral
leaned into reliance swimming
toward center
spinning birth
drift into the shape of your tomorrow
instead of being some certain thing
photographer
writer
husband
friend
allow yourself the possibility
of constellations
a tenuous thread of petal
to pine to starry surprise
embedded with the pulp and thistle
of your own happenstance
let memory's keepsakes
be moments of light
glistening through your time-
sculpted self
free from the weight
of weep and want and regret
the reed of your voice may whittle

toward solitude
a freefall that settles
in the loam of burrow and mulch
can you know your pining
is not for aught and this simmer
your primordial welter
can you trust that rain and earth's rust
will tender this
husk
will coddle your seed
from shelter?

AND THEN I FORGOT HOW TO WRITE A POEM

how to sit with the inconsequential
buttons and paperclips the rubber bands
and debris scuttled to the nether reaches
of drawers and glove compartments
orphaned keys and thumbtacks
milk-bone crumbs in the bottom of a purse
the Tupperware the shower rings
breath mints safety pins the random
unbidden memories
I don't know what to do with now

LAMPYRIDAE

World's fireflies threatened
by habitat loss and light pollution
—*Washington Post* headline

Even the dwindling darkness
becomes a commodity
relinquished to distant spaces
when dispelled by streetlamps.

Shadows recede into smaller shadows,
a reliquary shaped by fear.
And you, my luminescent lover,
inscribe eloquent calligraphies

in the troposphere of memory.
The filament of your fire drifts
along the retina soon to dissolve
in this chalky slate of sky.

YES

Sunlight slants into the reception hall
and glistens on bottle-shaped clusters of
butterfly bush flowers as the open-window
breeze nudges the blooms like my sister-in-
law's plumber of a brother who stands rigid
at the edge of the dance floor scowling at the
bride's gayest friends who shimmy like street
corner air-dancers inflatable tube men rippling
from the ground up as my sister-in-law's brother
resists the music rising from his feet though his
legs and torso surge like a windsock his arms
fused tightly against his sides hands shoved
deep in pockets as the hip DJ drops a heavy
bassline a thumpa-thumpa shockwave pretty
much trampolining all of us to our feet the
whole wedding party bobbing our bliss to
the beat and yes my sister-in-law's brother
wobbles his shoulders like jello and yes he
lifts and drops one heel now as the electro
fusion throbs the crowd a hypnotic tempo
and no his hands are still in his pockets
but yes he's watching the gay guys rumba
one behind the other their hands holding
the next boy's hips and yes the brother is
smiling he is smiling and yes it's a conga
line and yes music snakes along the dance
floor and yes his heel lands harder now and
yes the aunties rave in the queue and yes
we serpentine toward him oh yes the crowd
the music the butterflies yes oh yes oh yes

WHEN YOU FIRST CAME HOME

Little sack of potatoes, little Yukon Gold, I carried you with one hand supporting your neck, the other wrapped under you like a sling, elbow to fingertip. You had a language of sighs and hiccups, blinks and eyes following mine. *Hello!* little kitten, little head of tufts and black spun sugar. I dressed you, my boy doll, in your Levi jacket, your matching denim cap. A game we played. Bounce-bounce-bounce the baby. One, two, peek-a-boo. Did you know I lost twice before you? A game I didn't even know I was playing. But now Simon says kiss the baby. Simon says carry him gently. Simon says support his wobbly head and never ever let him go.

FOR IMMEDIATE RELEASE

The glaciers are going on tour.
NASA announced today:
a littoral circuit, coastal and global.
We'll flood your cities
with surges of music to last the ages.
Coming soon to a shoreline near you.
The tour lifts weight off the plates,
who are soon expected to announce
their own hard rock tour. Promoters predict
worldwide response will be tectonic,
shifting the poles of music trends everywhere.
From Norteño to Southern Rock,
Delta Blues to New Wave:
foundational changes launching
a tsunami of sound.
Get ready for the storm!
And save your seats now.

BREAKAWAY

A Steller's Jay caught his tail
in the fishing line we'd draped
over our thistle feeder
to foil the sparrows.

I grabbed kitchen shears
and tiptoed beneath the birches
where his mate scrawed
warnings of my approach.

What is a pendulum,
if not the desperate measure
of each remaining
moment we share?

The jay beat his wings,
thrashing against trunks and
clapboard till his tail-feathers tore,
leaving this unwanted talisman,

bloodied & wry. And isn't that
what freedom often requires:
the sacrifice of something essential
for a wingbeat's promise

of light, horizon, and sky.

A BOREAL CURE

Alder bark carries the stain of lychen
Blotched like ink splats in a Rorshach
Caldron where each spot bubbles your
Deepest fears, deepest regrets and who can
Even face these woods, their pulses and prompts, their
Feral grief, where leaf-fall blankets and buffers the
Gravelly layers of your mother, brother, sister, niece,
Haphazard and scattered, rained upon, leached
Into groundwater, shuttled to sea.
Just like the pillow, black top, car seat, where they were
Killed. Water washes everything, even this
Lingering melancholy that given time will
Migrate to unmitigated awe, the way shoots of
Nettle spire into splendor, upright in the light
Of greening, because grief is not five clean stages, but a
Plank, precarious as memory, not
Quite balanced, but enough to shore up these
Rickety legs. Nix the blessing, the
Silent prayer, the newly planted spruce.
Tomorrow sifts through the cedar branches
Under the Douglas fir. Fern fronds
Verify everything you believe about
Wan loneliness and its Fibonacci spread, the
Xylem of lost moments, the surety of
Yet more loss to come. Each visitation is its own blessing
Zinging toward us now, even as we speak.

EUPHORIC CHICKEN GLITTER

Rocking aisle 14 with the boneless
thighs and plumped up breasts,
my colorful neighbor, we'll call her Katie,
starts rolling her shoulders
and pumping her arms back and forth
like the former guy at a poorly attended
rally except there are no white people
wearing "Blacks for Trump" t-shirts
here. Anyway, Katie's gone full
disco, and the store speakers are feeding
a tinny version of Lady Gaga's "Born
This Way." Katie presses her pelvic bone
against the standup refrigerator case,
rows of plastic-wrapped chicken parts
her awestruck audience. *We're all born*
superstars, Katie shouts into the make-believe
microphone lifted to her glitter gloss mouth.
The guys behind the meat counter
pump their fists, too. *Right track, baby,*
they egg her on. She pulls out two
packages of bone-in thighs, probably
what she came for in the first place,
pretends to make-out with them, glitter lipstick
everywhere, and the Meat Department
Manager says, *Lady you're buying those*
and Katie presses one to her chest, the other
to that long gap where babies emerge
and prances toward the registers, dressed in meat
like Gaga herself, words trailing behind her,
something about being a queen, and I wish
I had half her nerve as I pop my head
in the frozen pizza case, trying to choose
between pepperoni and plain cheese.

LIGHT,
LEASH, MEMORY, NOW

FOR CAMPBELL, JUMPING OUT THE CAR WINDOW

he deciphers messages   like a paramour
hidden on the wind   longing for adventure
umbels of   sinews stretched
wild   when we drive
strumming   past the
last   dog park he whimpers for
ditch daisies   a romp
of late summer   loping with friends
seductions   over pocked turf   when
eyes scanning   he spots a doe grazing
roadside   near the gravel lane
and woods   he leaps out the window
where is the   sprinting
lover he never met   across a field of echoes
lingering under the   lost
howling   scent of
moon   pulse and pang

TERMINAL VELOCITY

The night after the world burned down
after sweeping winds brushed cinders
from the clouds and the sky exploded

    into stars for the first time in weeks
    we walked into earth's gravity well
    of grief still reeling from the news
    and argued as we reached the end

of our driveway where friends overheard us
on the way home from the candlelight vigil
for RBG and even though we could see

    the Milky Way and an occasional
    shooting star there was no wishing upon
    what with smoke still acrid in our lungs & eyes
    that shed yellow diamonds as if beauty

were not the jurisdiction of the past
as if we weren't all fleeting meteors
brief fireballs blazing in flight.

LIGHT, LEASH, MEMORY, NOW

The light laid a moving mosaic over the path
like sun confetti and the leaves glowed
a sorcery above us. Your laughter spread
across the canopy as though it, too, were light.

You held the dog leash in one hand,
my hand in the other. The crickets played
their hypnotic percussions and my thoughts wound
as much as the leash in the osoberry.

Memories lapped like the receding tide.
Some park on the sound, I've forgotten which
of the two men I thought I was in love with
at the time. Even the alders lost to memory.

Unlike today with these big leaf maples
and you and our daypack ready to open
into a picnic. Chocolates and cheeses.
A tangerine ripened between us now.

SO THERE WE WERE, TWO LOVE MONKEYS

Late-night walk up Sucia Drive.
Moon so bright our shadows
nearly talked back to us. Voices
gravelly as crows, and our hands
swung between us like a hammock
or the float of a chicken hawk
looking for home.

APOGEE OF APATHY

What if each window of the bus framed for us—
that is, each glass pane, fogged in winter
and the half-drowsed breaths of each passenger—
what if those glimpses, murky with strangers

on the street and the all-too-human
foibles of garbage and graffiti and grime
were recorded as pages in the books
of our lives? What if all the things we can't

unsee, the daily crimes, petty and otherwise,
maintained their equal importance, and couldn't
be erased by the biographers in our brains,
who handpick what others may or may not see.

What if our legacy were merely a litany
of our crimes against the earth, against humanity,
that we did not stop, that we watched
indifferently from the scratched and vandalized

windows of our daily excuse. What if
our children, and our children's children
had to answer for the drooping eyelid,
nodding stupor we bring to the world, the trash

we overlook when we walk the dog.
What if our only vestiges are the bones
of orphans and orcas, Shi'a and Sunnis,
we killed or let die on our watch.

FROM THE UPLIFT OF HEAT THERMALS

Evenings as the dishwasher gurgles and the dog dozes,
when you've done all the damage you can
for one day—taken on too much, worked

too hard, accomplished too little, and everyone knows
you're a disaster, with words like *amateurish, un-
professional, sloppy, severe, a mistake* echoing

in your mind—don't run from the disappointment you feel
for tomorrow you'll be walking the dog when a downy
woodpecker will call from the nest where she cradles

her eggs, where she might feel just as trapped and hapless
as you, but soon you will both look back from the uplift
of heat thermals, rising, soaring, as you take the sky.

FOR SKIPJACK, WHO SLEEPS EVEN NOW

We could never have been lovers—
you with your penchant for meat
and me savoring the slow melt
of chocolate,
my life a poison to you,
and yours, a never-ending marathon of naps.
You saw me as too unleashed,
even though I was always tethered
behind you.
Remember how you whined
when I wouldn't let you lick the water
from my legs after I showered?
My biggest complaint
besides your breath
was how you nosed your way into other
people's business and had to smell everything.
Still, we had fun racing down trails,
didn't we?
Your muscles rippled like a horse
pulling a chariot, and I followed behind
laughing as squirrels scampered up the trees.
You and me lovers?
No. But still we can't deny
we've always been in love.

WINTER DOG WALK

Woodsmoke drifts down the chimney, a specter
of weighted fog, and though its scent

coats my fingers from futzing in the woodstove
hours earlier, my mind knows the smell

emanates from a fire my mother laid
and father stoked in a cabin in Tahoma's

shadow. Flames danced on cedar and heat
billowed to the loft where my siblings and I

nestled under flannel quilts till the crackle
of bacon roused us into winter light.

Today the dog's leash lengthens and retracts
like memory. Smoke skims along roof shingles

and my thoughts burrow deep
in the frayed kip of a quilt.

NIGHT TRAIL

We ventured a level trail on our night hike,
hitched behind the dog. Our shadows
took long strides across the field. The beloved
described the twisted arc an arrow makes
when shot from a bow and how deer can bolt faster
than the arrow's feather-swifted flight. I wanted
to run his fingers along a sturdy-sleek tendon,
teach him the tension of hot-sprung require.
Clouds sequestered the moonlight. Hawthorn
and Western hemlock thickened along the trail's
vein wall. In the deepest thatch of darkness,
we held our breath for owl song. Tug
of leash, the dog smelled our way home.

ARS POETICA

Writing is like walking the dog. Sometimes it's leashed to form. A trail to follow. Or unleashed, roving, wild. We start as poet and transmogrify to animal, sniffing our way by impulse and urge.

Writing is an archeological dig. Or a hunt through the archives of memory and moment.

Writing is like falling asleep, surrendering to the delirium of dreams.

Writing means sculpting with sound. Part assonance, part consonance, and part dissonance. Part providence, arrogance, innocence.

Or forget the dog. Writing makes you a meerkat, screech owl, jellyfish, berserker.

Poeming breaks the bones of what we know. Resets them into a new scaffolding, a dizzying perch for perspective.

A poem is a ribbon of song. A knock on the door. An invitation away from loneliness. A dance into humility and hunger. Wave of slumber. Paeon for pain.

There is no how for writing. No when, or where, or why. Writing is wax on, wax off, inspiration, aspiration, nicker and neigh, lull and frenzy, flight and fray.

Writing mimics the journey of water, effluvial fingers seeping into an ocean, uplift of mist, downpour of sorrow, glisten of crystal, snowmelt of spring.

Writing is an act of survival. A release valve. The need to express. Lips pressed to mouthpiece or reed. A channel of sound shaped by embouchure.

Yet each word is an edict of hope. Whether vision or scree, aubade or nocturne. An incantation, lamentation, a cri de cœur, a plea.

A QUANTUM THEORY OF LOVE

STARGAZING

A normal star forms from a clump of dust and gas in a stellar nursery.

—NASA

I can't stop thinking
about dust motes, how,
waking from a nap,
I watched them, a galaxy
of golden stars drifting
across the late afternoon
of my living room.
That night
my mother-in-law texted:
Prognosis very bad.
The PET scan overlaid
with the CT scan, lit up
its own molecular cloud galaxy
of red giants, white dwarfs,
neutron stars: a universe expanding
through the lungs and lymph nodes
of her eldest daughter.
At any time,
the air can be laden with dust motes
and pollen, microbial yeast, mold
spores, and, yes, even star dust
so tiny they're impossible
to see with the naked eye.

ORIGIN STORY—GREAT BLUE HERON

after Todd Spalti's sculpture Tribute

We were born of sand and pebble on a cove in early spring.
We fledged in a nuzzle of feathers, opened our eyes to darkness.
We emerged from a tumble of loosestrife at marsh's edge.
We crocked our beaks to the beckoning sky.
Sticklebacks and sculpin slid down muggery gullets.
We waited at a linger of cordgrass to spear smelt and perch.
The wind leaned into us, rippled our glassine view.
We trained each eye.
One attended dart, the other followed swerve.
Our underbellies mimicked the weight of cloud.
Our legs stretched long into water.
They waggled behind us in flight.
We learned language in the crux of trees:
  chuck, wrick, and *vergue*.
By the summer of goldenrod we met our mates.
By autumn we waded again in solitary waters.
We cradled egg in our beak, that cracked-open world,
  as graytime growlers sketched shadows on the sea—
  balefire, rumble and quake.
Rock crab and spot shrimp scurried at our feet.
Tide pushed.
Tide pulled.
At night a slow stroke of wing ferried us home.

RAPTOR

Today in a flutter of leaves
a sharp-shinned hawk balanced
with one leg tucked under
his freckle-feathered chest
on a fence post
at the edge of our garden.

To get a closer look
I crawled through the living
room and onto the porch
crept to the ledge
and quietly lifted my head.

The hawk turned
razor eyes toward mine
with a gaze that sliced
open wonder
and startled that other
wild bird flapping
against the fenced cage
of my chest.

SLIPSILVER

Splitting rivulets from cool magma, we poured the watersilver like liquid metal mirrors swimming in the palm till we learned about absorption and the kidney's bitter end, how China's first emperor drank it mixed with powdered jade, his cheeks and fingers flaming as fast death replaced the glowing life his alchemists swore to extend. So when my toddler son bit through his glass thermometer I dialed 9-1-1 then did my best to clear his mouth and airway of glass shards and the elusive cinnabar. And when the medics swarmed over Jeff, I kept out of their way. The dispatcher looped in Poison Control, ran some basic tests, while the sirens drew neighbors, who whispered in the hall *Look how frightened Jeff is while his mother does fuckall.* Alchemy comes from Sanskrit: the way of mercury. My neighbors had wanted children. I just wanted Jeff to live. We weren't so different after all. An amalgam of base elements with every way to wound, yet nothing to forgive.

A QUANTUM THEORY OF LOVE

If all things with energy are attracted to one another,
    then my body, swaying to the light
as we stood admiring the Frida Kahlo self-portrait,

had simply found the nearest energetic field
    of your torso giving off heat
and the trace spoors of hunger in bloom.

Gravity being the weakest force in nature,
    physicists would tell us that attraction is basically
proportional to the product of our masses combined.

As though the space between your thigh and mine
    did not compress when some billion light years away
two black holes colliding sent a ripple

of gravitational disturbance that jiggered time and space
    until both thighs knew everything they needed
about mass and heat and unswerving magnetism.

TADPOLES

With our toes touching the lake's edge
we watched a simmer of frog eggs float
just beneath the undulating water.

Every day something new as the sacs
pulsed up from muddy dregs
just beneath the undulating water.

Mouths agog they siphoned great gulps
and swam the algae bed
where our toes touched the lake's edge.

Translucent tails shaped into pollywogs.
Their gills quivered. Their bodies spread
just beneath the undulating water.

What started as gelatinous globs
morphed and sprouted uncertain legs
where our toes touched the lake's edge.

Agog ourselves—me eleven, you twelve—
we barely believed our own legs could bud or fledge
but our toes touched at the lake's edge
just beneath the undulating water.

PI KE PUHT

For the Beloved

In Vashi he drank chai from clay cups
astringent with dirt. Thick, sweet,
and spicy like the dawn of his neck.
Sipped slowly, the tea spread tendrils
of root and pulse. The tossed cup shattered
to become one with the earth.

Chai-wallah, he cried
and reflexively twitched elbows and wrists.
Monsoon rains swept the silt of soil.
Its pliant clay pooled in the gutters
of his palms and collarbone.
Terra cotta smooth and sweltering.

He knew me by my broken shards,
recognized thirst
in the crumbling aftermath of goodbyes—
sun-dried, unglazed,
still fragrant with cardamom
and a ghost of sugar.

LOVE'S BLIND CONTOUR

When you said your face
formed a map of China,
I saw burled roots,

the tender of bark
folded on flexed arms, how
the musculature of muffled

ceiling clouds can skiff
past a cedar-hinged fall,
each hand a braille

on the reef side of morning,
each coast an eden,
after the squall.

TRAVEL JOURNAL

You sleep nine time zones west of here
and I imagine the cadence of your ribcage,
how it swells with the private air of our bedroom
then slowly shrinks back into itself.
I know
you will turn with a sound like a word that hints
of the dialogue in your dreams. *Plum* or *tong* or *yes*,
while here the espresso machine sputters in response.
The first month we were married,
I found your old
travel journal from some past trip to Paris—
your entry filled with mansard roofs, copper,
and the self-healing properties of zinc.
The next entry's script,
told of a hotel room romp,
jumping on the bed, someone's underwear
on someone else's head, and the squiggling legs
of those y's and q's written like a laughing heap on the bed.
Why didn't you tell me
you'd fallen in love
with Paris once before? Your answer
was to throw the travelogue onto the fire
and vow the rest of France is ours alone.
We never promised to talk or not talk
about past loves.
There are women who left you with the purple bruises
of a longing that never ends. And men who left me
with an aftertaste of iron that rises sometimes in a panic.
When we make love, or now,
in this quiet recollection
of intimacies, I know your voice
has been shaped by all the Parises of the world,
and mine molded in a convex mirror
where everything looms
closer than it appears.

ON THE ANNIVERSARY OF THE FIRST TIME WE KISSED

a great horned owl coohooed from the roofline
of an old log cabin. We had just emerged
from Mijita's, warmed by tequila and chile verde.
February and its star-spattered sky harkened

the years long passed since our earliest spring.
With each beckoning, the owl pitched forward,
lifting her tail feathers, sending lonely need
into char-glistened night. Another owl with a wan timbre

of wind, sang back to her. And we stood in the road,
two voyeurs watching the female raise her skirt,
the male rasp his wings, and both cry in passion and pain.
Or was the nocturne our own, a wide-eyed lilting,

tangled in vow and beseech, this unkempt music
of feather and talon, whistle and screech?

IN THIS LAND, EVERYTHING

is my mother. The mahonia berry
and dogwood blossom, swordfern
and fen. She taught me taxonomy,
showed me shape so I could find her
after her death, burgeoning
in trillium and salal,
tadpole and fawn lily,
bursting in a puffball of spores
everywhere, everywhere.

NAPTIME IN THE GARDEN

A bee nestles between the cool petals
of a dahlia, still but for the drift of antennae
wisping in the wind. Along the far path,
the edges of an ornamental maple ripen

in the sun. Clouds come and go. Their shadows
sweep the garden. Hollyhocks sway. The wheelbarrow
sits stout but empty. The neighbor's flag
riffles in the breeze. Sapsuckers tap into birch,

percussion for chickadee and finch. Now the cry
of waxwing. Now the sigh of leaves. And a memory
drifts through so gently one hardly notices.
Something about masks, or the election,

or was it that first pear on the last day of summer?
Apples litter the ground. Their scent wafts with rose petals
and wet humus, the last loam of summer
softening with sleep now, sodden with sleep.

BEDDED IN WET ROCK LIKE ANY OTHER ROE

Bedded in wet rock like any other roe,
I woke in a float at river's rim.

Sweet slurry of ancestors,
my pillow and pledge.

I dreamed down current,
a hatchling of spring's surprise.

Fry to smolt, I weltered
in fluvial estuaries and brackish tides.

When vast waters welcomed me,
our sweep of Chinook synchronized.

Outline of orcas. Shape of shark.
Such mercury molting.

We skirted the surf in a cross weave
of silver, bolted lightning back at the sky.

Rest wriggled in a forest of eelgrass.
Porphyra camouflaged our flash.

We swam hard against the waves—
building muscle, storing steam

—till the ancestors called us
from a spill of snow-fed waters.

Clear slipstream of memory.
River me home. River me home.

ABOUT THE AUTHOR

Jill McCabe Johnson lives and writes in Washington State's San Juan Islands, an archipelago in the Salish Sea. An avid hiker, Jill's research interests center on the influence of walking on the various literatures of the world and the power of the written word to inspire change. Jill's own writing, which doesn't shy away from difficult topics, leans toward hope, empowerment, and trust in the prevailing goodness of humanity.

Jill holds an MFA in Creative Writing from Pacific Lutheran University and a Ph.D. in English from the University of Nebraska—Lincoln. Honors include support from the National Endowment for the Humanities, Artist Trust, and Hedgebrook, among others. Jill is editor-in-chief of Wandering Aengus Press and its imprint, Trail to Table Press.

Publications include the poetry collections *Tangled in Vow & Beseech* (MoonPath, 2024), finalist for the Sally Albiso Poetry Award, *Revolutions We'd Hoped We'd Outgrown*

(Finishing Line, 2017), shortlisted for the Clara Johnson Award for Women's Literature, and *Diary of the One Swelling Sea* (MoonPath, 2013), winner of the Nautilus Silver Award in Poetry, plus the poetry chapbook *Pendulum* (Seven Kitchens, 2016), finalist for the Rane Arroyo Poetry Award, and the nonfiction chapbook *Borderlines* (Sweet, 2015).

Printed in the USA
CPSIA information can be obtained
at www.ICGtesting.com
CBHW021532050224
4055CB00005B/25